# Geologist Danny

written and photographed
by
Mia Coulton

I can see a big rock.

It is black and white.

I pick it up

and put it in my bucket.

I can see a red rock.

I pick it up

and put it in my bucket.

I can see a white rock.

I pick it up

and put it in my bucket.

Can you see all my rocks in my bucket?

I am on big rocks.

I am a scientist.

I am a geologist.